INSIDE THE KNOWLEDGE

- VOLUME 1 -

VICTOR MASOTTI

DEDICATED TO :

MY CAT – NANDO

MY MOTHER – LUCIA

[2] I – INFINITE OR FINITE?

[6] II – UNIVERSE'S SHAPE?

[8] IIa – OBSERVABLE UNIVERSE'S SHAPE?

[9] III – THE HISTORY OF THE MOON?

[13] IV – INTERNAL LAYERS OF THE MOON?

[14] V – VENUS TO BECOME THE NEXT EARTH?

[17] VI – WILL EARTH BECOME LIKE VENUS?

[20] VII – ABOUT INCREASING POPULATION?

[22] VIIa – WILL GLOBAL WARMING END US ALL?

[23] VIII – WHAT'S THE LIFECYCLE OF DWARFS?

[25] VIIIa – ARE BLACK HOLES SPECIAL?

[27] IX – WHAT WAS THE BIG BANG THEN?

[29] X – HOW MANY MOONS DOES SATURN HAVE?

[32] XI – WHAT ARE THE TYPES OF MOON?

[34] XII – DOES EARTH HAVE THREE MOONS?

[36] XIIa – WHAT'S THE KORDYLEWSKY CLOUD?

[37] XIIb – WHAT ARE STARS MADE OF?

[43] XIII – BLACK HOLES AND NEUTRON STARS?

[44] XIV – SHOOTING STARS AND ATMOSPHERE?

[46] XV – ANY SIMILARITIES BETWEEN PLANETS?

[53] XVI – WHAT WILL HAPPEN TO NEPTUNE?

[54] XVII – WHAT IS THE MILKY WAY?

[60] XVIII – WHAT'S THE ANDROMEDA GALAXY?

[67] XVIIIa – DO WE KNOW OTHER GALAXIES?

[70] XIX – WHAT EXACTLY ARE NEBULAE?

[75] XIXa – HOW MANY TYPES OF NEBULAE?

[79] XIXb – BUT ORION AND WESTBROOK?

[84] XX – HOW MANY CONSTELLATIONS EXIST?

[86] XXa – WHAT ARE ASTERISMS?

[87] XXI – WHAT ABOUT MUSLIM ASTRONOMY?

[89] XXII – THEN GREEK ASTRONOMERS?

[92] XXIII – WHO WAS NICOLAUS COPERNICUS?

[94] XXIV – DID THE ROMANS KNOW THE SAME?

[67] XXV – ANYTHING WE ACTUALLY MISSED?

PART I – INFINITE OR FINITE?

The problem that many encountered during their lifetime was about the simple question – is the infinite a thing or just some fake news that teachers tell us to temporarily stop our imagination.

I guess we will never know anything unless we brainstorm our thoughts and elaborate them into something pleasant.

Remember those geometry classes where professors would often tell you to make the lines seem infinite by adding some odd little segments at both ends? Pencils are a tool used by humans to sketch, design and as a cool piece of knowledge to put over your ear. So the answer to the question is determinable or not? Either we ask ourselves this, or we prove it in some way.

Infinity represents something that is considered to be boundless in some way or either endless. Often, people draw it as a turned-down eight or some could say it's a race track.

It really depends on your personality. Still, I believe it's an eight. However, since the time of the ancient Greeks, the nature of infinity was a massive subject of many, many discussions among philosophers.

The symbol was first introduced in the 17th century. Strangely, mathematicians thought this to be a new chapter for them, presenting the infinitesimal calculus.

Coming back to the line, if it is viewed as a set of points joined together, their infinite number is way bigger than the quantity of integers available. Infinity was also introduced in logic, computing, topology and further general analysis.

The first appearance of infinity could be argued to have been said by a Greek philosopher named Anaximander, who used the word "apeiron". It meant "unbounded" and other similar adjectives. You get the idea, this is not a new concept and it will never disappear. We aren't omnipotent to argue what reality means and we cannot come up one day with something totally new.

Greeks like Euclid stated the possibility of infinity existing by using examples such as straight lines and right angles. Their thesis were mainly based on the idea that it could be repeated as many times as possible. Here is the problem. Possible has an antonym that troubled our thinking from the origins. Is it probable? Is it going to happen or not?

The possibility of something happening can only occur when something causes it.

There are many mathematicians from around the globe that gave brilliant answers, such as Augustin-Louis Cauchy and Surya Prajnapti, to explain why infinity is possible. I guess that it can be possible, still we cannot be entirely sure.

If someone starts drawing a line on a paper, tell me how is it going to be infinite. The paper might end, it might rip, the writer's hands could exhaust, the line could be drawn inaccurately... - It is not infinite by many ways. The concept of infinite needs a better definition. Is it mathematical, philosophical or logical? The same idea applies to matter. Without thinking in one of those ways, it is literally impossible to come up with an answer.

I admire Antoine Lavoisier, as he taught us that in his Law of Conservation of Mass, "matter is neither created nor destroyed... the total amount of mass and energy in the universe is constant. It is difficult to point a problem because we simply are not gods with maximal qualities flying around the vacuums, discovering the unknown. The law was stated in 1785, and since then, it has not been defeated. Is there a reason for this? Is it because our minds aren't able to think that far ahead?

In fact, the volume is determined and changed constantly. As an example, black holes are formed when too much 'information' is combined together. Obviously, it's harder to explain but this is the simple way of defining it. They are created by matter and evolve in a different matter in appearance.

At this point, we are feeling confused because it's just frustrating for humans to be uncertain. This applies also to the idea that death is unpredictable, and leads our souls to continuously repeat ourselves this: "when am I going to die?... And how?" I do not know that for now.

Have you ever seen something endless? Videogames make us think that we can play as much as we want. Programs, like everything, are like another way of representing storage. Websites and other are simply hard drives full of terabytes and petabytes filling up each time we do something. This means that we are changing something. Therefore, this is not endless.

Change disagrees massively with Infinity because if something like a segment is going and going in the same direction, it can be stopped or reflected or moved. This makes it stopping. Some will argue that the line is still going in some cases, and that's true. However, the line can be stopped in many other ways. Pause and think about it for a second… or two – better three.

If change is unpredictable and part of our life, we cannot determine if it will last forever or not. A website without something behind it, will disappear.

Maybe, infinity describes only some things and not actions. Probably, it could be that everything we do is forever-lasting.

Philosophically speaking, everything might be registered somewhere, and even if humanity will extinct, our story will still be found. Personally, I believe in what I see, at least, I prefer to see reality like this.

Our minds are created to focus on what matters to us, and to repeatedly think and imagine the world in those ways. Also our minds are not endless. We will die and forever lose that precious information. Our souls might be around, as people say, but there is no evidence that our thoughts are carried around like books and that we can grab them whenever we need them. I mean that when something ends, it just ends. That's it.

Following the previous statement, if everything ends, the concept of things being "endless" is illogical. Am I right? What's your opinion? It may be that our emotions affect our reasoning because we don't want to believe for example that we will die, but if it is certain, why than we even talk about it.

Let's face it. Everything ends. The sun will explode, our phones will die, our plants will stop growing one day, our cars will brake… and so on. Therefore, believing that infinity exists is partially wrong because we just proved that everything we know can end or change in some way.

However, we can't be certain of matter's behaviour. As already said, we are humans and not omniscient gods. What we see is what we know and what we talk about. Isn't it easier to write a paper about ducks rather than the universe, galaxies and black holes?

In conclusion, the term was founded, like the Bible, a long time ago, and because of this, things might have changed or ended, and our minds developed – at a point – where we consider ourselves differently. Obviously we are not able to tell it, but I contend that we have developed so much, positively or negatively.

PART II – UNIVERSE'S SHAPE?

The shape of the universe is the local and global geometry of the universe. The features of it are mainly described by its curvature. However, in the topology of the universe describes general global properties of its shape as of a continuous object.

The spatial curvature is related to general relativity, which is there to describe how spacetime is curved and bent by both mass and energy, whereas in the spatial topology it cannot be determined from its curvature. Those spaces with different topologies exist mathematically.

In addition, spacetime is literally any mathematical model which fuses the three dimensions of space and the one dimension of time into a single, four-dimensional manifold. Those types of diagrams are used to visualize relativistic effects, for example why different observers perceive in different ways where and when the events occur.

Recently, cosmologists distinguish the observable universe with the entire universe, as the former being a sort of spherical portion of the latter that can be accessible during observations.

Using the Cosmological Principle – which is the notion that the spatial distribution of matter in the universe is homogeneous and isotropic when's viewed in a large enough scale – the universe that we can observe is similar for all vantage points.

As a bonus, the universe can be described in three simple ways. The first being finite or infinite, as we already discussed. The others two are its curvature position and the connectivity of how it is all put together into one.

We are able to determine that a universe with a positive curvature – same as saying closed – is necessarily finite.

The exact shape of it, is still a matter of debate in cosmology. However, from the sources we currently have in the first half of the 21th century, like WMAP, Boomerang and Planck, it is said that the universe is nearly flat, having a 0.4% margin of error in the calculations.

PART IIa – OBSERVABLE UNIVERSE'S SHAPE

The two aspects to consider when talking about the shape of the observable universe are its local geometry – which concerns the curvature of the universe – and its global geometry – which concerns the topology of the universe as a whole.

It is thought that the universe can be thought of a square that extends outwards from any observation point for about 46.5 billion light years.

In fact, it is long (9.461km x10^12)x46.500.000.000 light years. This calculates how long the universe is in km.

It is 4.399.365.00E+23 km wide.

Logically, if the observable universe encompasses the entire universe, we could be able to determine the structure of the whole universe only by observation. Isn't it shocking news? However, if it is smaller than the entire universe, our observations would be limited to an extremely small corner of the universe. I mean, we just saw how wide it is.

Imagine how many milky ways you could fit in there. Is like filling an ocean with thin grains of sand.

Back to the point, it is improbable we will ever know everything, giving that our lifetime on Earth could end way before the explosion of the Sun or a bigger star.

After some long reasoning, we find out that we cannot tell the shape of the universe at a point in time. We might not think about it, but time is a key factor when talking about this. That's why we have light years to represent over 9 quadrillion guitars next to each other.

PART III – THE HISTORY OF THE MOON?

The origin of the Moon is usually explained by a Mars-sized body striking the Earth, making a debris ring that eventually collected into a single natural satellite, the Moon, but there are a number of variations on this giant-impact hypothesis, as well as alternative explanations, and research continues into how the Moon came to be.

Other proposed scenarios include captured body, fission, formed together (condensation theory, Synestia), planetesimal collisions (formed from asteroid-like bodies), and collision theories.

Some theories have been stated that presume the proto-Earth had no large moons early in the formation of the Solar System, 4.425 billion years ago, Earth being basically rock and lava.

Theia, an early protoplanet the size of Mars, hit Earth in such a way that it ejected a considerable amount of material away from Earth. Some proportion of these ejecta escaped into space, but the rest consolidated into a single spherical body in orbit about Earth, creating the Moon.

The hypothesis requires a collision between a proto-Earth about 90% of the diameter of present Earth, and another body the diameter of Mars (half of the terrestrial diameter and a tenth of its mass).

The latter has sometimes been referred to as Theia, the name of the mother of Selene, the Moon goddess in Greek mythology.

This size ratio is needed in order for the resulting system to have sufficient angular momentum to match the current orbital configuration. Such an impact would have put enough material into orbit around Earth to have eventually accumulated to form the Moon.

The newly formed Moon orbited at about one-tenth the distance that it does today, and spiralled outward because of tidal friction transferring angular momentum from the rotations of both bodies to the Moon's orbital motion.

Along the way, the Moon's rotation became tidally locked to Earth, so that one side of the Moon continually faces toward Earth.

Also, the Moon would have collided with and incorporated any small pre-existing satellites of Earth, which would have shared the Earth's composition, including isotopic abundances.

The geology of the Moon has since been more independent of the Earth.

A 2012 study on the depletion of zinc isotopes on the Moon supported the giant-impact origin for Earth and the Moon.

To help resolve these problems, a new theory published in 2012 posits that two bodies—each five times the size of Mars, collided then recollided, forming a large disc of mixed debris that eventually formed Earth and the Moon.

The paper was called "Forming a Moon with an Earth-like composition via a Giant Impact".

There are also many other types of hypothesis of how the Moon came to be, which include the merger of two planets, multiple impacts, the Moon being captured by the Earth, via fission, accretion and nuclear explosion.

The easiest to understand of the bunch is the fission, explaining how the Earth was spinning so fast that a piece of mass was expelled. This was proposed by George Darwin in 1879. Other people, like the Austrian researcher Otto Ampherer suggested the possibility that the Moon emerged as a cause of the continental drift, when the continents moved across the ocean bed.

The hypothesis of the accretion explains that both the Moon and the Earth formed together as a double system from the primordial accretion disk, a structure formed by diffuse material in orbital motion around a massive central body, or a black hole.

However, most do not support this due to the fact that the Moon has a relatively small iron core compared to the Earth, implying one was different than the other by a lot.

Lastly, another well discussed option was introduced recently by Dutch scientists Rob de Meijer and Wim van Westrenen in 2010 that the Moon was formed from a nuclear explosion caused by the centrifugal force of an earlier and spinning proto-Earth.

The centrifugal force would have concentrated heavy elements such as thorium and uranium on the equatorial plane and at the boundary between the Earth's outer core and mantle.

If the concentrations of these radioactive elements were high enough, this could have led to a nuclear chain reaction that became supercritical, causing a nuclear explosion ejecting the Moon into orbit.

PART IV – INTERNAL LAYERS OF THE MOON?

The Moon is a differentiated world. This means that it is composed of different layers with different compositions.

The heaviest materials have sunken down into the Moon's centre, and the lightest materials have risen to the outermost layer.

Seismic, rotational, and gravity measurement studies have allowed us to gain insights into the different layers within the Moon.

At the centre is the Moon's dense, metallic core. The core is largely composed of iron and some nickel. The inner core is a solid mass about 480 km in diameter.

Surrounding the solid inner core is a fluid outer core, that brings the total diameter of the core to about 660 km.

The Moon's core is smaller (about 20% of the Moons diameter) compared to other terrestrial worlds like the Earth, with cores measuring closer to 50% of their diameters.

Above the core are the mantle and crust.

Differences in compositions between these layers tell a story of the Moon being largely, or even completely, composed of a great ocean of magma in its very early history.

As the magma ocean began to cool, crystals began to form within the magma. Crystals of denser mantle minerals, such as olivine and pyroxene sank down to the bottom of the ocean.

Lighter minerals, notably anorthosite plagioclase feldspar, crystalized and floated to the surface to form the Moon's crust. The mantle, with a thickness of roughly 1350 km is far more extensive than the crust, which has an average thickness of about 50 km.

Did you know, the crust of the Moon seems to be thinner on the side of the Moon facing the Earth, and thicker on the side facing away. Researchers are still working to determine why this might be.

PART V – VENUS TO BECOME THE NEXT EARTH?

In the future events that humans will escape Earth to search for a new planet, choices for now are going to be restricted to our solar system.

Many are speculating Mars, however it's essentially impossible to live there with those extremely hot temperatures.

Yet, no one has ever really considered and invested in going to Venus. Why not then, due to having nearly the same mass and size as Earth.

Unfortunately, Venus' atmosphere is entirely composed of CO2, carbon dioxide, which is incredibly poisonous to our body and we cannot stand it for much time. In addition, the atmospheric pressure is ninety times more compared to Earth.

And some clouds appearing on top, are made of sulphuric acid, which is one of the main three types of acids found and very toxic for any living condition. It can destroy ecosystems, nature and living beings.

Also, sulphuric acid gives Venus a very shiny look. This is actually great news because the atmosphere being so thick, lets through only three percent of the sunlight.

Despite the lack of sunlight and temperature caused by the sun, Venus is very hot, that could melt many materials, with degrees reaching over 370 degrees Celsius. We cannot live with this ridiculous high temperatures.

However, talking about daily events, Venus has a very rare and particular rotation in the solar system. The sun rises in the west and ending in the east. Also, being very slow, a single year on Venus lasts 225 Earth days and a day lasts 243 Earth days.

Also, Venus has no moons and has already been visited by many spacecraft.

Essentially, point being made that considering Venus for now is not the greatest option is we don't invest some resistant materials and machines to go and live there. We would need to form new ecosystems.

And if you wanted another confirmation, it's like going to hell. However, think about the things I told you before, isn't that a reflection of Earth once upon a time. I mean, Venus probably was a stunning planet back in the days. Probably had water and oceans and tectonic plates were still moving.

When this happened, four and a half billion years ago, the Sun was particularly smaller. Stars start small, grow big and expand light, and then end with explosions. So, as the Sun aged, Venus became hotter and hotter that living beings started to disappear. Probably, some could be found down inside the surface!

When oceans started to evaporate quickly, water vapour rose up in the atmosphere. And this same vapour was a very great assistant at trapping heat there.

Venus essentially become like the future of Earth, a hot greenhouse with any species dying due to extreme heat exposures. And without oceans and acting in any form, the plates just started to stop and remain still.

The real problem, was carbon that just like gasses trapped in ice here on Earth, was also trapped in the dirt, which slowly, outgassed and escaped.

So, at the end of the argument, is it every imaginable to live with no compromises on Venus?

Recently, astronomers could actually have found some signature of life on that toxic planet. Indeed, evidence from researchers founds phosphine, which's a gas associated with living organisms, and is very present in the only habitable region of Venus' atmosphere that we know of.

Even though we already know that Earth and Venus are like twins, Venus is very easy to deduct as a very challenging planet to ever live on.

However, researchers have found that in a space in between found in the atmosphere, temperatures can range from thirty degrees to two hundreds, talking Fahrenheit.

PART VI – WILL EARTH BECOME LIKE VENUS?

We just stated that Venus was probably a nice place to live in, and probably had thousands of ecosystems depending on its previous characteristics, temperatures and nature.

Yet the Venus we approximately know today is worse than living in hell. So why are we making this point?

Earth is the twin of Venus in some ways, and this is actually terrifying. If Venus, became hell, could this happen to our planet too, and how long must we wait.

We all know the Earth will catastrophically end, and hundreds or more ways have been found in which it could end. For example, global warming could kill any living being or the Sun being a star could explode and destroy the solar system.

The question is then, before jumping to conclusions, will Earth become like Venus, will it stay the same until the end or will it end much sooner than we thought at the start?

Some quick facts to note down, are for example the Sun that is becoming brighter by ten percent in about every billion years from now, causing our planet's temperature to rise dramatically high, reaching one day forty-seven degrees.

This means briefly, that just like Venus, Earth's oceans will evaporate and it will become a greenhouse planet. Indeed, every decade the Sun will rise the temperatures, by even just a bit.

Many say that it is totally unlikely that Earth will become like Venus anytime soon, also because a billion years is just a bit too much for us to even understand.

Some say that in seven and a half billion years, the Earth will be absorbed by the Sun. But what certainly won't happen, is for the Sun to become a huge black hole. Being a star, the Sun is just not as big in mass to become a black hole.

Instead, what will happen to the big star, is the removal of all its explosive and hot layers, and become a bright white dwarf planet. It will be still incredibly hot, but will not produce any nuclear power.

Another quick fact, is that Stars have three fates. One is again, white dwarf, another's is the black hole, which isn't an option for the Sun, and the last one being a neutron star. A star's fate is really determined by its mass.

Before returning to other fatal ending possibilities of our planet, let's explore why the Sun will not become a black dwarf. At least it might, but white is the very first stage of a dwarf.

Black is the last and as far as we know, it could take up to trillions of years for that to occur.

As just spoken, Earth will still stick around for a long time, yet that's not the same for all the species on Earth.

Indeed, all plant and animal life on Earth need oxygen to survive.

According to a new study, a billion years from now, Earth's oxygen will become depleted in a span of about ten thousand years, bringing about worldwide extinction for all except microbes.

Something actually cool to know and probably the end of either Earth, or any similar planet in this galaxy, is that every star will be replaced and no matter the size become a black dwarf. As explained, black dwarf can take a lot of time, depending on the planet's mass.

So, this will probably occur in more or less a hundred trillions years from now. We cannot be certain, but it is very likely based on what we have and know by now.

Also, Earth and Venus and the others will follow that trend, but the time it will take only depends on the density of the universe and galaxy.

PART VII – ABOUT INCREASING POPULATION?

In 1960, the population of the only planet where life has no limits, for now, had only 3 billion people.

So imagine, we cannot know during the course of time how many died, if we counted right or any similar problem, if in two thousands and so much more years our population reached that number, how did it outgrow in just less than a hundred years?

Indeed, as of the start of the second decade of this new millennium, in 2021, our population will reach eight billions humans beings, probably by 2030 or even 2023 or 24!

With this lightning-fast rate of births, counting deaths in some circumstances, our population will reach its first two digit number and the other nine zeros, by 2100, the first century of this millennium!

But wait, isn't the Earth the size of the Earth, and nota as huge as the Pistol Star, which is a hundred times bigger in size and ten million times brighter compared to the Sun we know, how is the population going to fit?

Will we actually really need to escape Earth once and for all?

Well, we actually are not very sure, but what we know if that in just a century, our population has nearly tripled and that initially to reach the first billion, we took over two million years.

Something interesting to know, is that the seven billionth child to exist, was from India and born at seven and twenty-five minutes am of the local time, in Mali village, found in Uttar Pradesh state.

Now, Earth could sustain our lives for one to four billion years. It is unlikely, but still possible. In fact, some labs support the idea that human lives have a ninety-five percent chance of extinguishing in less than eight million years. Probably, already half of human's history has passed.

PART VIIa – WILL GLOBAL WARMING END US ALL?

By many students, Earth is considered to be effected by Global Warming in the future.

However, we know that I twill bring high temperatures and this could cause the human extinction, or possibly not.

Global warming is considered to be a natural effect, as the Sun will become hotter, and also because there is no way around it.

In fact, we are not causing the global warming, rather we are just speeding up the whole process.

If the world, were to become to hot, it would destroy essential-to-life ecosystems, and determine our end.

So the question becomes again, can we stop it?

While we cannot stop global warming overnight, or even over the next several decades, we can slow the rate and limit the amount of global warming by reducing human emissions of heat-trapping gases and soot.

Yet this doesn't mean our planet will not become a greenhouse sphere. It will, but it'll take much more time. In fact, by now we should all know the possible lifespan of the Earth in the future. Not certain, but still is an option to look out for.

The hottest place for now, in the world, is not in Africa or Asia, but in the well-known Death Valley. At the Furnace Greek Ranch, on the tenth July of 1913, it was recorder to have passed fifty-six degrees Celsius.

As of now, the ten hottest places in the world are Death Valley in USA, Kebili in Tunisia, Mitri bah in Kuwait, Turbat in Pakistan, Dallol in Ethiopia, Aziziya in

Libya, Wadi Halfa in Sudan, Dasht-e Loot and Bandar-e Mahshahr in Iran and Ghadames also in Libya.

Hopefully got them all right!

PART VIII – WHAT'S THE LIFECYCLE OF DWARFS?

Dwarf stars all have a fate, which as I told you only depends on their mass and size. They normally all or most tend to start from a Stellar Nebula, and like the Sun, if they are not big enough, they'll become an average star.

After that, they might become a red giant, transform lately into a planetary nebula, and finally enter the first stage of a dwarf, which is white.

Otherwise, being bigger or even huge, they could become a massive star, than a red supergiant star and mutate into a Supernova.

Two choices will come then at the door; either become a neutron star by clearing off any dust around it or explode massively, creating a very large black hole.

Neutron stars are very dense and compact stars thought to actually be composed mainly by neutrons. They are not big at all, with a twenty kilometres of diameter, but can be extremely heavy, even though weight is not an issue in most parts of the universe.

Also, they can be very dangerous because they have huge and strong fields. If a single neutron star entered the solar system we live in, it could throw orbits of the planets around like toys, and also could rip planets apart, due to the enormous force caused.

Yet, nothing to worry, as the first neutron star that is nearer to us is more than five hundred light years distant!

Lastly, just to finish off this neutron argument, these stars can actually be destroyed and become also a black hole, or stay nearly the same for a long period of time.

Remember that, if two of those were to meet, they could form a giant force and a giant black hole on their own.

PART VIIIa – ARE BLACK HOLES SPECIAL?

Black holes essentially are regions of spacetime where gravity is so powerful that literally nothing can escape it at any time, even light being so fast. And there are a lot of them, actually they're really common.

A huge amount of mass normally can cause a black hole. Before moving on, just know the four types of black holes; there are stellar, intermediate, supermassive and even miniature.

Even though they are black and we think they are holes, actually they simply are not producing any light and are very big and wide.

Also, they aren't empty. A lot of matter is condensed into a single point.

The good news in this deep sea of fear, is that supermassive black holes, being so big have a stronger gravity force but a much weaker stretching force. In fact, it's just like stretching a cookie before baking it, which is small and can be very fast stretched. A pizza can be harder to stretch because it just takes more time.

Mediocre analogy, but you get the point. In fact, you could actually survive for some time in these black holes.

A cool and horrifying thing to know is that a black hole as small as a rubber, could cause the end of the world, as they just get bigger every time they eat.

Remember that studies show that time, just like sound and weight in space can be very misleading. Time's extremely slow near a black hole, and it could make you feel like time has stopped.

Yet, not all black holes are lucky. If they are not big enough with enough mass, they could suffer from hawking radiation, reducing even more their smaller mass, and theories say they could vaporise.

So, shrinking is a pretty valid option.

Let's just say that some of the smallest might take over three billion years to evaporate and some even more than the lifespan of the universe.

Also, as I told you before, it could take about twenty minutes or less and the whole Earth could be sucked up by a two centimetres black hole.

In addition, we talked about the nearest neutron star, but the nearest supermassive black hole is about two thousand light years away, which is not so far.

But the closest, is actually called the Unicorn, with fifteen hundred light years far from us. In our galaxy, black holes are mostly invisible rather than black, and very hard to spot in some cases.

In this universe, before anything existed, another supermassive black hole converted itself into a "white" hole, and became the Big Bang. Also, even though it could be considered a "coloured" hole, it did not form any black hole.

Mass and energy was spread equally across, with no margins of error that could cause a collapse.

PART IX – WHAT WAS THE BIG BANG THEN?

The big bang is also believed to have started in a very odd yet obvious way. It is said that an enormous amount of energy was jam packed into a tiny point. Also, remember that science has no typical definition of size.

However, this very dense point exploded quickly with unimaginable force to create also the billions of galaxies there are. Pretty titanic explosion!

The Big Bang is actually been proven with a glow in the Universe called Cosmic Microwave Background.

When the Universe began and expanded, the explosive light stretched into microwaves. Some microwave telescopes can still see this light. This microscopes show the whole sky filled with a glow, no matter when.

Now, as said, everything has an end, same for the whole Universe.

If it holds too much matter, including dark matter, the combined gravitational attraction of everything that exists will slowly halt the expansion and suddenly collapse. Over time, galaxies and stars and planets will smash each other, one by one, killing any living being and probably form many black holes.

Before going on, we must all acknowledge that dark matter is essentially matter composed of many particles which are immune to light, due to not reflecting it, absorbing it or emitting it at all.

Also, black matter cannot be detected by electromagnetic radiation, as again is immune to light or its properties. Just like some black holes, we know they exist because of their effects.

The Universe will expand forever, until it lasts. While, the galaxies will merge together with groups and clusters, all the massive galaxies will move away from each other.

All the stars then will die or otherwise get sucked by supermassive and giant black holes. All the stellar corpses and massive matter bodies will collapse or from even bigger black holes.

We are not sure if this is the future and how long it will take to occur.

Also, the Big Bang is said to have given the start to everything we know. In addition, it created many universes, which now are known to be separated by extremely long distances.

Essentially one universe contains many universes. The Universe we know could be called by many as the Multiverse, and also then called the Cosmological Multiverse.

PART X – HOW MANY MOONS DOES SATURN HAVE?

As we all know, some planets or dwarfs have moons orbiting around them. Venus and Mercury have zero moons.

Earth has one moon, simply called the Moon, but others have even more. Mars is the second in line with two.

However, there are the big boys that have many more. Neptune has fourteen, Uranus has twenty-seven, Jupiter has seventy-nine and Saturn has eighty-two.

Now, both these two last planets mentioned only have fifty-three moons confirmed, but still Saturn is said to have more.

Before moving on, moons are essentially natural satellites, and can be very different from each other by size, type and also shape. Generally they are all solid and a few even have atmospheres.

Most of them were formed by discs of gas and dust circulating around the planets.

Remember that there are hundreds of moons in this solar system. Moons with a proper name are considered to be moons at a hundred percent, while some with different properties or that are harder-to-find, are only given a name with a starting letter and a year.

Some cool facts are for example how the Moon was formed. A large body as big as Mars crashed onto Earth, ejecting a lot of material in space and reducing the mass of the now-called Moon. This is said to have occurred about four and a half billion years ago.

Also, we tend to think that moons are some kind of perfect spheres, but they aren't all like that. The two moons of Mars, Phobos and Deimos, are not similar. They have the same moving and circulating habits as the Moon, but are very dark and lumpy.

Phobos is moving towards Mars and could collide within the next fifty million years, or it could break due to gravity and create a thing ring around Mars.

Now let's move to the very important moons. Even though Saturn is said to have more moons, Jupiter has the biggest one, Ganymede. It also has an ocean moon, Europa, and a volcanic moon, Io.

Many of Jupiter's moons have very high elliptical orbits and orbit anticlockwise.

Saturn, Uranus and Neptune have also very irregular moons, with even similar properties.

Now, let's talk about Saturn, which we know as having eighty-two moons orbiting around it for now. Saturn has two ocean moons, called Enceladus and Titan.

Both contain subsurface oceans, and Titan even has surface seas, which are composed of ethane and methane. You could probably swim in those seas, and not risk to burn with Methane as air isn't that present there.

In fact, Titan's atmosphere is made of nitrogen and methane, and a bit of hydrogen.

Lastly, Titan is also the second biggest moon in the solar system.

We are going to talk more about these types of moons, but first let's explore some others.

Uranus' inner moons seem to be made of half water ice and half hard rock. Miranda is one of them and very unusual. It's body shows scars of impacts from very large rocky bodies.

And, Neptune has a moon called Triton, which is actually as big as Pluto, and orbits too backwards.

Pluto is now not anymore a planet, yet a dwarf. It still has moons, and its largest is Charon, which is just half the size. Its formation is very similar to our Moon.

Yet, the trend here seems that the closest to the Sun, the less moons you will have, like Ceres, a dwarf that has zero.

PART XI – WHAT ARE THE TYPES OF MOON?

We previously discussed and explored which are the main moons. And some can be volcanic, rocky or ocean.

There are some that have volcanic activities, and they are Io, Triton and Enceladus.

The bodies, and not moons, with volcanoes are Earth and also Venus. Now, also the Moon is said to had volcanoes many million or billion years ago.

Jupiter probably had some, but now seem to have not. Mars actually has the Mount Olympus, which is the biggest volcano in the solar system. And it was found to be still live, which very early eruptions, occurred just twenty-five million years ago or less.

Let's recap. Venus has many small volcanoes spread on its surface, Io is the most active moon in the solar system and the Moon probably has many volcanoes too.

Ocean moons also exist. The most famous and obvious is Europa, which is made of silicate rock and has a water ice crust with an iron-nickel core inside. The characteristics of this moon could lead to the theory that a water ocean is found under the surface, just like in Antarctica.

Some might think that is water is present we could live there, but it isn't always true. It has a very thin oxygen atmosphere, which is too little for us to breathe. Also, Europa has very strong magnetic field to shield against the deadly radiations coming from Jupiter's surface.

Then the question is, could you live on Titan or Enceladus. Titan is very similar to Earth, and is stated to be like the moon with in terms of the necessary requirements to support life.

Indeed, it could be the most hospitable extra-terrestrial world within the solar system, apart from Earth.

Now, we know that it has more nitrogen compared to Earth, and very few oxygen or hydrogen.

The main reason and problem why life would be hard there is not only breathing, it's also temperature.

Titan's far colder than Earth, with some reaching -179.6 degrees Celsius. Now that is the maximum it can get, but with billions of years or less, it might become hotter and open up its oceans.

Before losing our hopes, the minimum temperature is much more habitable, with only three and a half degrees, which is very cold for us, but much more doable compared to the maximum.

PART XII – DOES EARTH HAVE THREE MOONS?

Even if you live in Antarctica or in the jungle, you probably have been told that there's a Moon orbiting our planet. We have been on the Moon and took samples.

Now, we know everything about eclipses, the colour, shape, and whatever's left, yet no one has wondered if our Moon had brothers.

Indeed, a team of Hungarian and other Polish researchers, astronomers and physicists found out over 50 years ago that Earth has two extra dust moons surrounding it and orbiting the planet.

The first glimpse of this discovery was made by Polish astronomer Kazimierz Kordylewski, after whom the "dust cloud" were named.

These dust moons resulted to be huge in size, yet less visible due to being composed of many tiny dust particles that barely measured one micrometre in diameter.

When the Sun hits those particles, they start glowing and since these natural satellites emit a faint light, they're very difficult to find in the sky, which sunny days, stars and other covering them.

These dust clouds are always changing and actually have existed for a very long time, talking about millions of years.

Indeed, they are very stable when orbiting, but the dust particles which are tiny, swap out and interchange, so they are always different in mass and maybe colour or light properties.

How did this Polish researcher find their approximate position. He used Lagrange points which are spots in a planetary orbit where the pull of gravity that's working from two opposing celestial bodies, is balanced due to the force of their orbits.

Hard to understand, but you get the idea. In the 1950s, Kordylewski searched two points, L4 and L5, where he found the first glimpse of the dust bodies.

Another question would be, are they useful or dangerous? Well, they might help us to better understand fuel consumption and other safety issues of space.

As we already know these dust clouds are not as stable in terms of particles and mass, unlike other similar bodies like comets or meteors, and so on.

These particles yet stuck together to form a dusty planet, which occurs with the Lagrange balancing act. So even though Kordylewski Dust Cloud is a strange body, is might have been there for a very long time, hence why it's considered a moon.

PART XIIa – WHAT'S THE KORDYLEWSKY CLOUD?

Yes, the Kordylewski Dust Cloud has been a moon for a very long time, and just to recap, these clouds are very large concentrations of dust particles, that exist at the L4 and L5 points of the Lagrangian points of the Earth-Moon system.

They have been confirmed recently, just some years ago in October of 2018.

For now, we don't know much, and the part where we have explored the most is indeed on their appearance.

Their brightness can be compared to the gegenschein, which is a faintly bright spot in the bright night sky, centred at the antisolar point. The backscatter of the sunlight caused by the interplanetary dust actually causes an optical phenomenon, also called counterglow.

PART XIIb – WHAT ARE STARS MADE OF?

We now know the brightness of Kordylewski Dust Clouds to be faint and hard to spot, due to the similarity compared to other smaller bodies.

Yet what are stars and other "rocks" actually made of? Stars can differ a lot in brightness and size, yet they are very large celestial bodies made mostly of hydrogen and helium gas that produce light and heat from the nuclear reactions happening inside their cores.

Now, this seems a lot like a Sun situation, as other smaller stars may be a bit different.

However, moving on with this theology, what happens is that these stars get very hot. Indeed, the core has very high temperatures and pressure.

What happens then is that four protons fuse to form more helium, in many steps. This process releases huge amounts of energy, and make the stars shine.

As stars age, what happens inside them changes.

As we just talked about, we can tell stars are not solid, yet a hot ball of gas made up mostly of, yes you named it, hydrogen.

Stars as huge as the Sun and beyond are actually made up of plasma, the fourth state of matter.

Now, plasma is very interesting as no one ever told you of its existence with accurate details, am I right?

So, it's called the fourth state of matter coming after solid, liquid and gas.

Essentially, you could say it's a rebranded version of gas. In this state of matter, an ionized substance becomes highly electrically conductive to the point that long-range electric and magnetic fields are present and key in their behaviour.

Now, plasma is fluid, like a "liquid mixed with gas", but has very high electrical charges.

Because the Sun is so hot the gas and hydrogen were heated up to create plasma, as it is made of nearly hundred percent of plasma.

Actually, did you know that the Sun is white, and emits all the colours of the rainbow? And that it emits all the types of wavelengths?

Getting back to stars, let's talk about temperatures. White stars are hotter than red and yellow, yet blue stars are hotter than any of them.

And, obviously, all stars die. For example, the Sun will end in about five and a half billion years from now, as nuclear fusions occur quite often on the Sun, hydrogen acts to help speed up, while shrinking and shining more.

One day, the Sun will run out of hydrogen and began expanding by burning helium.

This means it's going to become a red giant and as already told you explode to become a white dwarf.

The sun consumes about five million tons of its nuclear hydrogen every second, and that's a lot.

The lifespan of a star totally depends on its size. The Sun will die a lot quicker than much smaller stars.

Why? Because in space, we prove constantly that physics are clueless of what's happening.

Essentially, if the Sun takes about ten to five billion years to ran out of hydrogen, stars as big as supernova can die in just a tenth of a billion years.

You guessed it then, a star can be really small. And the smallest are called red dwarfs. These can take up to a hundred billion years to explode.

This is because the reactions take so much time to complete, and can be even older than the universe.

As you understood by know, there are a lot of types of stars here and there. Stars are classified with a Spectral Classification or Code system.

Before telling you the types, let's define some key information. The temperatures of the stars tells u show bright it is.

And the pressure tells us their approximate size.

Stars are classified with "classes". M stands for red stars, which are the coolest and smallest of all. They measure about two thousand to nearly four thousand kelvin. This is the surface temperature.

Then we find orange stars, starting with the maximum temperatures of red stars and maxing out at five thousand and two hundred kelvin. These are named K.

Yellow stars are called G, and they reach the six thousand kelvin.

And there is a yellowish white type, called F, that limits at seven thousand and a half kelvin.

Known as A, white stars are big and hot. They start from the three quarters of the five digit number and end at ten thousand kelvin.

The last two are similar in colour but has a very different approach to temperature.

White blueish stars called B have the longest range of temperature we know, starting at ten thousand kelvin and maxing out at thirty-three thousand. That's more than a lot.

However, blue stars get even hotter with no defined limit and classified as O.

To remember these, you could use a mnemonic sentence, like "Oh Boy, An F Grade Kills Me."

Even though the Sun is white in space, it's classified to be a G type, with temperature as high as six thousand kelvin, nothing compared to blue stars.

The nearest blue star to our Sun is Eta Carinae, located about and seven and a half light years away.

This blue hypergiant star is a hundred times bigger in mass compared to the Sun. Think about it as the Sun being a grain of sand and Eta Carinae as a large football.

Before moving on to any other similar topic, let's talk about world records.

In our Milky Way, there are many huge stars. Wolf-Rayet stars are found to be four in this galaxy and six in other external galaxies.

The hottest of all, is surely WR 102, of spectral classification WO2, and measured just about two hundred and ten thousand kelvin. So the take the Sun in comparison, is just about three thousand and four hundred percent hotter, you know, nothing much.

The coldest star then, well we know something about that.

Researchers from the Penn State University found out years ago that the coldest star is WISE J0855-0714 or WISE J085510.

How cold it is? Well, just about two hundred kelvin and maxing out at nearly three hundred approximately.

Yeah, it's as cold as going to the remote places of Alaska, and I mean where it's very cold, being just about negative seventy degrees Celsius.

Stars also come in very different sizes.

Take neutron stars, they can be just about twenty to forty kilometres in diameter. However, white dwarfs can be similar to Earth's.

On the other hand, the largest of all measure about fifteen hundred times the diameter of the Sun, and we are talking about supergiants.

The Sun, being a star, is a hundred and nine times bigger than the Earth in diameter, and surely hotter.

So, we should have a very good knowledge about stars now, also because there's lot to know about them, unlike dust moons.

PART XIII – BLACK HOLES AND NEUTRON STARS?

We know that continuous collisions occur arguably every day in the universe. But what happens when a neutron star collides with a black hole.

We know how black holes work approximately.

But, let's revise what neutron stars are. Essentially they are compact stars mainly composed of neutrons. Typically, they are pretty small in size, as many cities are even wider than those stars, in diameter.

Also, they indeed are made only about a ninety-five percent of neutrons, and protons found convert to neutrons, ending up to releasing ubiquitous particles known as neutrinos.

So, what happens then, well the situation is sort of complex in some ways. So, let's suppose in this case the black hole is larger than the neutron star.

What occurs and has been seen in many occasions, is that they form a spiral, like the ones in the ocean. Once they'll meet at the centre, the black hole will quickly swallow the neutron star.

PART XIV – SHOOTING STARS AND ATMOSPHERE?

Shooting stars, also known as meteors, are caused by tiny parts of dust from space. These burn up to sixty to a hundred and thirty kilometres per hour above Earth's surface, as they fall at very high speeds into our atmosphere.

To sum up, a small rock moves fast, so it heats up quick and glows through our atmosphere. So these are called meteors.

Let's first define the difference between a meteor and meteorite. A meteor is a dust rock that falls on Earth, however, it burns and heats up, so crumbles and does not reach the ground.

Now, the few meteors that don't explode in air and reach the planet on the ground, become meteorites.

Also, these are not even stars. Yes, because there is another difference, and is between falling star or shooting star and a normal star.

Shooting stars are amazing streaks of light that are sometimes seen during the night. We call these stars because they shine, but really that's just particles breaking through our atmosphere.

In fact, when you fall through the atmosphere of the Earth, the break-ness speeds needed to break into it, generates a lot of pressure, a lot more than seas'.

And, let's analyse the layers of the atmosphere that gives us life. First we start with the Troposphere.

It's the very first and located low, starting off from the ground up to ten or twelve kilometres high. Commercial planes don't usually fly that high due to the atmosphere becoming thinner and thinner as you go up. It's all about safety issues.

So, what is the Troposphere? It's thick at the equator and makes up for eighty percent of the mass of the whole atmosphere. It's also where nearly all weather conditions take place.

Further up, there's the Stratosphere. This second layer maxes out at fifty kilometres from sea level. Obviously, temperatures decrease with height, ozone layer is found in between this one. And, it makes up for twenty percent approximately of the mass of the atmosphere.

This means that some birds can fly that high, and some are swans, cranes or even vultures.

Now, there's also the Mesosphere, that extends up to ninety kilometres. And, is the coldest of the bunch, as it freezes water vapour into clouds very fast. Carbon dioxide is a key factor.

The two big boys come next. The Thermosphere is the hottest and the second thickest layer of all, going up to eight hundred kilometres high.

Satellites and space stations are found within. It is indeed the hottest because of the "lack" of molecules and atoms, the air temperature still rises.

The last is the Exosphere. As we all know, meso stands for middle, thermo for hot and exo means external, indeed being the outer layer.

The air there is thin, mostly made up of hydrogen and helium. Others are also found, like oxygen.

Also, it's made of nearly nothing in terms of mass and sound waves stand no chance to pass through, as space is a vacuum and lets no sound in.

Also, being there would set you back good, freezing cold in seconds.

PART XV – ANY SIMILARITIES BETWEEN PLANETS?

In this case, we are talking about similarities not related to the shape or size, but if their inner structure, cores and layers are similar or very different.

Every inner planet has a core, made with different materials but all with the same mission; to stick around and orbit around the Sun.

Let's then explore each planet and the Sun, one by one.

The Sun, is like any other else, is a ball of gas. Talking numbers, it has a high concentration of hydrogen, with a ninety-one percent complex and has a nine percent helium.

In terms of mass, the Sun is just about seventy percent hydrogen and thirty percent helium. So it's switched up a bit.

Now, we did not talk about layers of the Sun, but that's for the upcoming future.

Mercury is the smallest planet in the solar system. Just like our planet, Mercury's core is composed of liquid metal, but we do not know if some of the inner core it truly solid. Some have found the evidence that is indeed solid, and nearly the same size as Earth's.

Mercury has similar layers, like all. In terms of the biggest and most important, it has a mantle, than a really thick outer molten core and a minor solid inner core.

Mercury's surface is also like our Moon, with a dusty and rocky upper surface, containing craters caused by space rock impacts.

Its inner core is made of iron and the outer crust is again, rocky material. Not a surprise there.

Venus, a very interesting planet, has a crust that's significantly older than Earth's constantly changing surface. Venus has a "thin" crust probably made of rock.

It's mantle is way thicker than the core, unlike what happened with Mercury. Now, we cannot be extremely sure, so it might be wrong, partially.

We also are not sure if the inner core is solid, so we cannot confirm much about that.

Also, as an extra, Venus' magnetic field is much weaker than ours, because it's closer to the Sun, meaning it needs less force to remain orbiting.

Earth. We should all have learnt our structure. Literally, everyone who had a minimum of education in whatever way should.

However, we know Earth better, having solid and liquid cores. Earth's structure very easy to understand.

It has layers which approximately are all the same thickness, approximately alright.

There is the crust, the upper and lower mantle, the liquid outer core and the solid inner core. Now, we know even more, talking about the top layers.

Indeed, just like the something spheres we had in the atmosphere, we have them in the ground and below.

The lithosphere is the rocky outer part of Earth. Between the two mantles there is a minor transition zone and the mesosphere, which indeed is in the middle. Last, the endospheric part at the bottom.

In addition to those, we also find the asthenosphere, that is the zone lying beneath the lithosphere and is much hotter and fluid than that outer layer. It is involved in the tectonic plate movement and other adjustments. Very important to know.

And before moving on to Mars, as we all know the Moon's and no one really cares about that rock, there are the two types of crusts. Please tell me you know them by heart.

These two are the oceanic and continental crusts, which "make up" the surface we stand on.

The oceanic crust is made of dark basalt rocks, rich in minerals and important substances. The continental crust is made up of granite rocks and full of oxygen and silicon. Oceanic crust is in the oceans and continental is found above sea level, normally as it depends for every territory.

Essentially, continental's low in density and oceanic has a high density. Continental's thicker, floats on magma and cannot recycle whereas for the oceanic is the opposite of these three factors. So, just grab a geography book in this case.

Finally, let's move onto Mars. Yes.

Before starting, we already know Mars has the biggest volcano, which is the same size as Arizona in the USA. Literally that wide and big. I know.

And we know it's very active, yet slow due to its very large size, with the volcano measuring approximately six hundred and twenty-four kilometres in diameter.

So Mars is internally composed of iron, nickel and sulphur on the inside, pretty similar to Earth.

Mars is smaller, but still has approximately the same basaltic crust in terms of width. Has a pretty substantial mantle, when related to its structure.

And, there is the core, as always, but we are unconscious wherever it's solid or liquid. In fact, Mars has a basaltic crust, just like the Moon.

Jupiter is a fascinating one. It has clouds on top, a gaseous hydrogen inside, some liquid hydrogen, some metallic hydrogen and a core, which compared to its size is very tiny.

So to sum up this bit, it has an outer and inner atmosphere, with hydrogen and helium found on both.

After that there is a very thick and huge, like huge, fluid layer, made of the metallic hydrogen and helium, and you know that.

The inner core, is pure solid rock. Made up of surprisingly iron and frozen water, yeah H2O.

So the core is essentially dense yet hard rock.

After Jupiter, we find Saturn. So, it has again an iron and nickel core, dense and full of those metals. It's surrounded by rocky material and other compounds solidified by the heat and pressure.

Saturn's enveloped by liquid metallic hydrogen inside a layer of liquid hydrogen. Similar to Jupiter's core but much smaller.

On the outside, these planets are similar. Saturn has an organic-rich atmosphere and surface. An outer shell made of frozen water and other substances.

Underneath there's a global subsurface ocean, A high pressure ice shell and inside a hydrous silicate core, which we already talked about before.

Uranus instead of having a rocky molten core, it contains icy materials. The liquid core makes up the eighty percent of the mass of the planet, mostly made of water, methane, and ammonia ice.

Like other planets, it has a ring system, with a unique configuration, and it has a magnetosphere and numerous moons.

On the inner side, it has an outer atmosphere, with an upper cloud layer. In the inside atmosphere, we find helium, hydrogen and methane gases.

We have a huge mantle, made with ammonia water and methane ices, and at the centre there's the core. It is made of silicate, iron and nickel rock.

And, remember that Uranus is blue. Why is that? That's because of methane, found in its hydrogen and helium atmosphere. This planet is often dubbed as an ice giant, since it's made of, you named it, eighty percent of "water". A fluid mix essentially.

Of course, here's Neptune. Again, even more bluish than Uranus.

However, it has an outer atmosphere composed of hydrogen, helium and methane in gas form. And because it is blue who would have guessed these "ingredients".

Still, the inner atmosphere contains the same gases, but methane is a bit different and there are some minor changes.

After that, like with Uranus, we have a mantle. However, this mantle is more a frozen huge layer found within. It's made of water, ammonia and methane, and all in a frozen form let's say.

The inner core is not solid, at all. It's pure solid rock and frozen water mixed together.

So, as you can imagine, these planets at the end of the system, are freezing cold, like the opposite of hell combined with the opposite of relaxation.

Now, these conditions might be true as long as mankind will exist, but is very possible that they could either stay the same or become hot planets, meaning Earth will be a "secondary Sun", in that case.

Also, cool fact. Neptune has winds that would move you about two thousand kilometres per hour, so if you need to get out of town, here's how.

So it would be impossible to stand on Neptune, because it's freaking freezing, but also because of the wind we just mentioned.

PART XVI – WHAT WILL HAPPEN TO NEPTUNE?

Yes, it will be a short "lesson". We know the Sun is just a tiny yellow G-type star, and that is going to be as big as a red giant. Not fun.

So, we know that all these planets won't exist when that will occur, but the real question in mind is whenever the furthest will survive.

Neptune could be described in three words; wind, ice, water. So if the Sun would reach this planet, would the Sun become the largest oven we know?

Yes, Neptune will become an ice cream. It has volatile rings made of all the gas giants, contains icy compounds and will not experience a good ending.

Neptune will melt, boil or even sublimate when the red giant will begin its expansion. Lots of different state changes, right?

So, let's restart and understand what a red giant is. It will all start in five billion years. The Sun will initiate the process by burning helium away, turning into a red giant, so a big star, increasing temperatures.

It's like when you are a small kid and own a little plastic red car, and after twenty years you're driving a red Ferrari.

PART XVII – WHAT IS THE MILKY WAY?

So, we know the Milky Way is the galaxy in which we live in. And is a spiral shaped galaxy which contains many hundred billion stars, including our Sun.

It has a diameter about a hundred thousand light years long, and ten thousand thick.

So, this huge galaxy contains over two or four hundred billion stars. It also includes gas and dust, as well as a lot of dark matter.

Remember theirs is an order of sizes. The universe contains galaxies, and galaxies contain the solar systems. These contain stars, planets are smaller than stars and moons are smaller than planets. And asteroids are smaller than moons.

The term Milky Way comes from Latin, Via Lactea which means Milky Circle.

The Milky Way was observed since the Greeks and further defined as having many bodies by Galileo Galilei in 1610. Until the early 1920s, many thought that this galaxy contained all the stars in the universe.

But, that would mean this galaxy was the universe, so the universe was the Milky Way.

After many other discoveries, we found out that there are many other galaxies far away.

We just talked about the diameter of this barred spiral galaxy. Recent simulations are suggesting that a disk of dark matter, containing visible stars, may extend with a diameter up to two million light years.

Now, the Milky Way has several satellite galaxies, which form part of the Virgo Supercluster. This itself is a component of the Laniakea Supercluster.

This Laniakea Supercluster is as huge as huge, containing the Milky Way and over hundred galaxies, so how many trillion of stars will be there?

In the Milky Way, you find the Galactic Centre, which is the rotational centre, essentially a supermassive black hole with a size of about four million solar masses, and don't worry, is very far away from us.

Stars and gases appear at a wide range of distances from that centre, still going up to two hundred and twenty kilometres per second, so it's fast.

This speed discovery is contradicting the laws made by the Keplerian Dynamics, which were published in the 1600s, describing wrong the reality.

Now, it was the 1600s, you probably would be killed if you said there was another Sun.

So, these laws suggested that the mass of the Milky Way is complicated, as we can only see ten percent of the whole galaxy. Making the rest invisible.

This conjectural mass has been termed dark matter, as we know.

In addition, there are many distances related to the galaxy; yet the most fascinating is how fast it moves as a whole. It's six hundred kilometres per second.

And, if we want to know how old it is, well we have some news. Some say the Milky Way is as old as the universe itself, forming just after the Big Bang.

Let's talk about some very interesting facts.

For example, I told you how many stars there are in the Milky Way, but being sort of invisible and a huge number, simply imaginable, we are actually not sure.

Some say a hundred and others more than four hundreds.

For as much as we try, we simply have no precise idea of the mass and weight of the Milky Way. Estimates say it's from seven hundred billion to two trillion times the Sun's mass.

As said before, many say that approximately ninety percent of the mass is dark invisible matter.

And some incredible technology led to discover than our galaxy probably is living in the middle of an empty spot of the universe. It's like floating.

Another one you say? Well, many astronomers are constantly trying to photograph the supermassive black hole at the centre of the Milky Way.

And even if the Sun is like dust compared to that black giant, we simply have not photographed it.

As always, we know that many bodies were created or shaped due to heavy collisions, which happen every second.

Our Milky Way moves as a whole and sometimes crashes with other similar galaxies.

Obviously, some of these are tiny compared to ours, and orbit casually around. Some have been called Small or Large Magellanic Clouds.

The name's due to the Portuguese explorer Ferdinand Magellan, who understood this theory back in the 1600s.

You may also want to know that this Milky Way is very full of toxic grease. This means there is some dirty oily organic molecules that make up some bunches which move around our galaxy.

Now, these are scientifically known as aliphatic carbon compounds, and they are being produced in certain types of stars.

A recent study found that these substances could account for a quarter of the Milky Way's interstellar carbon. And remember, carbon is essential for many things.

We know what is going to happen to our star in many billion years and we talked of hundreds of events that are going to happen.

But, exactly what will happen first. This event will determine every other event. And probably it might be this one.

It might erase everything I told you here, and everything could be considered a lie.

So? The Milky Way might crash with its close neighbour in just four billion years. This neighbour is the Andromeda galaxy, and the Milky Way is moving towards that galaxy at about four hundred thousand kilometres per hour. So it's far away, but not that much apparently.

And what is even more sad, is that we'll be the victims of the collision. As the Andromeda galaxy is massive and bigger, it would swallow up our own galaxy and survive with probably no problem.

Did you know that mysterious bubbles in our galaxy are arising out? Imagine in your house, there has always been a monkey, which you never saw and one day you did, yeah.

In 2010, scientists discovered uncovered gigantic structures stretching for twenty-five thousand light years. These are named the Fermi bubbles.

And something to notice, is that these bodies emit gamma-rays. So how did they form.

About six to nine million years ago, our supermassive black hole ate a huge clamp of gas and dust, which made it out as glowing "radioactive" clouds.

And last quick and main fact, is that our Milky Way is being bombarded by bizarre and odd energy pulses from some parts of the universe.

This phenomenon was recorded vastly over the last decade. Essentially, they are fast radio bursts, or FRBs, which are mysterious signals shooting flashes of light from distant cosmos.

And we have zero idea why that occurs. We have captured from thirty to fifty FRBs, yet believe me when I say, these are not like finding a needle in the middle of a sand castle, it's way harder.

PART XVIII – WHAT'S THE ANDROMEDA GALAXY?

We live in this huge galaxy that moves so fast and is as big as big can get in this galaxy, but we are not alone.

I mean, we don't know other species of humans, but we are certain there are other galaxies far away. The Andromeda galaxy is very interesting, as we know it's going to be swallowing our bodies in a few years.

So, we need to know more. The Andromeda galaxy is also known as Messier 31, M31, or NGC 224 and originally named Andromeda Nebula.

This is obviously a barred spiral galaxy approximately two and a half million years away from Earth.

The galaxy's name stems from the area of Earth's sky in which it appears, the constellation of Andromeda.

The name itself comes from the Phoenician, not Ethiopian, Princess who was the wife of Perseus in the Greek mythology.

As we know, determining the mass of a galaxy is not an easy job, yet we can estimate that the mass of the Andromeda is about twenty-five to fifty percent larger than ours.

If this is true, we would be swallowed, but how can we be sure of anything that will happen in even a million years ago.

Indeed, it could be smaller.

Other estimates indicating the number of stars suggest a number similar to one trillion.

So, we need to talk more of what could happen if the stats were real.

Our galaxies will collide in about four to five billion years, and yeah we know it.

So, before going I know the ending, let me tell you about some good or maybe bad news.

Either our planet and solar system gets sucked up by the supermassive black hole, or it will collide to from a giant elliptical galaxy, or maybe a large lenticular galaxy.

What are those? An elliptical galaxy is a type of galaxy with an ellipsoidal shape and a smooth featureless image, not like our that look like disks.

These were first leaked and presented in a book by Edwin Hubble, and is one of the three classes of galaxies.

Now, a lenticular galaxy is another type of galaxy intermediate between the elliptical and spiral. Yes, the third "class". It has a very long and large disk, but does not have any things surrounding it, like a large spiral.

Another crispy fact, is that the Andromeda galaxy is among the brightest of the Messier objects, making it visible by our eyes, naked, even when the sky is not clear.

Before moving on, let's understand the Messier objects. This is a set of hundred and ten objects catalogued by the French astronomer Charles Messier.

These were discovered and saved because Messier was only interested in finding comets. Because of that, he listed every single body he was frustrated by.

It's like trying to read something on your average phone when the Sun is on you.

We know that the first to address the existence of our galaxy, the Milky Way, were the Ancient Greeks.

But who discovered the Andromeda galaxy? It was a Persian astronomer called Abd- al-Rahman al-Sufi, who was the first to discover it and describe it. This was in 964.

Later on, in 1612, the German astronomer, Simon Marius, gave an early look and description of the Andromeda via telescopic observations.

In 1745, Pierre Louis Maupertuis conjectured that the blurry spot was an island universe. In 1764, Charles Messier catalogued Andromeda as object M31, as you know and credited Marius for the discovery. However, the first was the Persian astronomer and was incorrect because it can be viewed by our naked eye.

In 1785, I know - long history, the astronomer William Herschel noted a faint reddish hue in the core region of Andromeda. He believe that that galaxy was the nearest great nebulae, and based on the colour and magnitude, he incorrectly guessed that is was not as far.

After that, in 1850, William Parsons made the first drawing of Andromeda's spiral structure. Years later, in 1864, Sir William Huggins noted that the spectrum of Andromeda differed from that of a gaseous nebula.

The spectra of the foreign galaxy displays a continuum of frequencies, superimposed with dark absorption lines, to help identify the chemical composition of an object.

Indeed, to make it simple, it's very similar to the spectra of individual stars, suggesting a stellar nature.

In 1885, a supernova known as S Andromedae was seen, the first and only saw by humans for now.

That was after named Nova 1885. In 1888, it doesn't stop, Isaac Roberts took the first picture of the Andromeda galaxy, which was thought to be a nebula within our Milky Way. So, let's celebrate for the picture and nothing else!

In 1912, Vesto Slipher used a spectroscopy to measure the radial velocity with respect to our solar system. And it was the largest velocity yet measured at three hundred kilometres per second.

In 1917, Heber Curtis observed a nova within Andromeda. Eleven more novae, after, were discovered and on average ten magnitude fainter than those that occurred elsewhere.

He came up with some results, and joined the island universes hypothesis, which held that the Andromeda galaxy "nebulae" was an independent galaxy. And sure he was right.

After that, in 1920 the Great Debate between Harlow Sharpley and Curtis took place. Yes, they have debates, I know. That was the Twitter of a hundred years ago.

It was setup to talk about the nature of the Milky Way, spiral nebulae and the dimensions of the overall universe.

In order to support his claim, he noted the appearance of dark lanes within Andromeda, which resembled the dust clouds of the Milky Way, as well as historical observations of Andromeda's Doppler shift.

And much more. After that, in 1922, Ernst Öpik presented a new method to estimate the distance of Andromeda using the measured velocities of its stars.

He did very good calculations. But in 1925, Hubble concluded the debate, because they last five years apparently.

So, it was ended by identifying the extragalactic Cepheid variable stars for the first time ever on astronomical photos of Andromeda.

These photos were taken on the 100 inches Hooker Telescope, which enabled to know the exact distance.

His measurement finally determined that Andromeda was an external individual galaxy. Yes!

And in 1943, Walter Baade was the first human on Earth to resolve stars in the central region of the galaxy. He identified two distinct populations of stars based upon their metallicity, naming those two young and high-velocity stars in the disk Type I and the older, yet red stars in the bulge Type II.

This nomenclature was subsequently adopter for stars within our Milky Way, and also elsewhere.

Baade also found out that there were two types of Cepheid variable stars, as well as the remainder of the universe.

In 1950, radio emission from the Andromeda galaxy was detected by Hanbury Brown and Cyril Hazard at Jodrell Bank Observatory. The first radio maps were made later in that decade by John Baldwin, and other collaborators, which stated that the core of the galaxy is called 2C 56 in the 2C radio astronomy catalogue.

In 2009, the first planet may have been found in the Andromeda. And those my friend are huge news, for real.

So, it was discovered with the technique of microlensing, which is caused by the deflection of light by a massive object.

Yet, we have no idea about it, but we gave it the name HIP 13044 b.

PART XVIIIa – DO WE KNOW OTHER GALAXIES?

We do know others, even though Andromeda might be the most know and visible.

So let's go through each one.

Again, there is the Milky Way and the Andromeda Galaxy. Ours is located 0 light years away, because we calculate ly, or light years, starting from our galaxy.

Andromeda is two million and a half ly away.

In between, there is LMC, which is a satellite galaxy, meaning it's tiny.

And we already know what is it. Remember Ferdinando Magellan, well this is the Large Magellanic Cloud, and classified as a Magellanic spiral.

Before going on, please know that there are hundreds, and I will list the most important initially.

Moving on, the Cigar galaxy, very far away with eleven and a half million ly. This galaxy is a starburst, part of the Ursa Major constellation and discovered by Johann Elert Bode in 1774, in Germany.

It has an elliptical shape produced by the oblique tilt of its starry disk relative to our line of sight. Maybe he was just having a cigar during the discovery.

And last detail, its five times more luminous than our Milky Way and not even talking about the centre.

Next in line of the big boys, there is the Pinwheel galaxy, and wow do they have fantasy when naming in Science in general. It's about twenty-one million ly away and it is unique, not having a black hole at its centre for now, as far as we know.

But, where it does not lack is the sources of x-rays. These sources are stellar-mass black holes, which are normally formed when stars die and their material falls into other black holes, or other types, it depends.

And it is also known as Galaxy M101, or called a flocculent spiral.

The Sombrero galaxy, please end this naming now! So this galaxy, named after the typical Mexican hat, as they always celebrate, is also known as Object 104, M104 or NGC 4594.

And this is the way we name professionally each galaxy or object in the universe.

So, the Sombrero galaxy is a spiral galaxy in the constellation of Virgo and Corvus, so in between.

The hallmark of the nearly edge-on galaxy is a brilliant, white, bulbous core encircled by thick dust lanes, which make up the shape. The centre is thought to be home to a "ultra" massive black hole.

And next up, Whirlpool galaxy. I guess they are running out of synonyms by now.

It's best known as M51, where the M stands for Messier. Like any other galaxy, it has a big supermassive black hole at the centre, surrounded by rings of dust. The core is very active, making this a Seyfert galaxy.

And guess what, it's not alone, as it has a friend around it called M51b and is a dwarf galaxy, which we will talk about in just a moment.

That black hole emits very strong powerful x-rays.

After that, the NGC 1300, I knew they would ran out. Obviously, it's named like this for many other scientific reasons.

This galaxy is situated over seventy million ly away, on the banks of the constellation Eridanus.

We don't know much, but we know about Tadpole galaxy, damn it they have done it again. So this is a disrupted barred galaxy located over four hundred thousand million ly from Earth, in the constellation Draco.

Its name comes from the tadpole shape, with the shape and star formation resembling that, the tadpole.

And it does have a long tail, actually very fascinating.

Lastly, there is Hoag's Object, around six hundred thousand ly away. Hoag's Object is a peculiar galaxy, consisting of a luminous central core, surrounded by a ring of gas and dust, where new and young blue stars are constantly forming.

This is a ring galaxy. Named after Arthur Hoag in 1950. And it does not have many stars, only about eight billion, which is nothing compared to Andromeda.

Remember that many galaxies are contained in the Hubble deep Field, which has thousands of galaxies, with all the ones we know and the furthest.

PART XIX – WHAT EXACTLY ARE NEBULAE?

Nebulae are enormous interstellar clouds of dust and gas which occupy the space between the stars. They sort of act like nurseries for new stars.

The roots of the word nebula come from Latin, meaning a mist, vapour, fog, smoke, exhalation. Indeed, nebulae are made up of dust, basic elements such as hydrogen and other ionized gases.

This term was originally used to describe any astronomical object, including galaxies.

Most nebulae are of vast size, and some are hundreds of ly in diameter.

The history of seeing nebulae has existed for a long time. Around 150 AD, Ptolemy, an astronomer, recorded in his books of his Almagest, five stars that appeared nebulous.

Then, he noted a region of nebulosity in between the constellation Ursa Major and Leo. This was the very first nebula and mentioned first by the Persian astronomer Abd al-Rahman al-Sufi, which we already know about.

So, we know he sort of discovered the Andromeda galaxy, but he also pointed a little cloud nearby.

The Omicron Velorum was catalogues by him in his books as a star cluster, and as a nebulous star, and also others like Brocchi's cluster.

Before continuing, an interesting fact; in 1054, Arabic and Chinese astronomers observed SN 1054, the supernova that formed the Crab nebula.

So, before going into too much details about specific nebulae, let's see the history of discoveries. In fact, the next subchapter will be all about the Orion nebula.

To being with then, in 1610, Nicolas-Claude Fabri de Peiresc discovered the Orion nebula using a simple telescope. Why, because it's the brightest nebula we know so far.

And in 1618, it was also observed by Johaan Baptist Cysat. However, the first detailed study ever came in 1659 from Christiaan Huygens.

A century after 1610, in 1715, Edmond Halley published a list of six nebulae. This number continued to increase as the years passed.

For example, Jean-Philippe de Cheseaux, a Swiss astronomer, added another twenty nebulae in 1746. And from 1751 to '53, Nicolas-Louis de Lacaille catalogued even more. He added forty-two, most of which were totally previously unknown.

In 1781, as we know, Messier stepped in and made his objects list, compiling over a hundred and more.

The Herschel family really did the job. William Herschel and his sister Caroline, who was also an astronomer, catalogued over a thousand new nebulae, and at the time it was huge.

That occurred in 1786, but they did not stop there. Three years later another thousand were found and in 1802, another half a thousand appeared.

The Herschel really did revolutionise the field of astronomy at the time. They also found out more back in 1790.

In addition, William Huggins in 1864 examined a particular spectra of seventy nebulae. He noticed that approximately a third of them were releasing an emission spectrum of gas.

The rest showed a continuous spectrum and thus were thought to consist of a mass of stars.

And the third category was established in 1912 when Vesto Slipher showed that the spectrum of the nebula that surrounded the star Merope matched the spectra of the Pleiades open cluster.

Thus the nebula radiates by reflected star light.

Now, nebulae before being discovered, they need to form in some way.

The are many mechanisms and each for a different type of nebulae. Some of those simply form from gas that's already in the interstellar medium.

Some form from stars. Star-forming regions are a class of emission nebula that's associated with giant molecular clouds.

These actually from as a molecular cloud, which collapses under its own weight, which produces stars.

The process is partially complicated, still. Essentially, massive stars from in the centre, with their UV radiation ionizing the nearby gases, making this visible at optical wavelengths.

That region of ionized hydrogen gas surrounds massive stars, and known as the H II region in science.

While the shells of neutral hydrogen surrounding that region, are known as the photo dissociation region.

Examples of this kind, so the star-forming regions, are the Rosette nebula, Omega nebula and Orion nebula.

We will explore those more after. For now, let's see the result of supernova explosions. The death of massive short-lived stars scatters material, which are again, thrown off to be ionized by the energy. Also, the object created by its core helps the process too.

The Crab nebula is the best example, found in the constellation of Taurus.

That event occurred in 1054, when the compact object was created, just after the explosion, it lied in the centre of the Crab nebula.

Even though there are many methods, there is another main one; to from as a planetary nebulae.

This is actually the final stage of a low-mass star's life, like our Sun. Stars that have a mass up to eight or ten solar masses, turn into red giants, slowly losing their outer layers during pulsations in their atmosphere.

What happens? Well, when a star loses enough material, the overall temperature increases and uv radiation is emitted to ionize the nebula that was thrown away.

Essentially, the end of our Sun is simple. It will become a red giant, explode to become a tiny white dwarf, create a huge surrounding nebula.

Again, if we are going to be eaten, we don't know what will happen, as we might be greater, different, or just painfully die. Who knows.

PART XIXa – HOW MANY TYPES OF NEBULAE?

We already talked about the main types, but let's go through them again with slightly more detail.

So, starting off easy, there are the classical types. Normally, we classify nowadays nebulae into four major groups.

However, before being fully understood, galaxies or spiral nebulae and star clusters were too distant to be seen as stars, so were classified as nebulae.

This is not happening anymore, and all have been divided.

The first being diffuse nebulae. These are very extended and stretched, meaning they have no well-defined boundaries.

This type is subdivided into other three subcategories. Emission nebulae, reflection nebulae and dark nebulae.

The ones that we see mean that they are visible light nebulae, and divided into emission nebulae, that emit spectral line radiation from ionized gas, like hydrogen.

As we know, normally known as region H II.

Also, keep in mind that reflection nebulae are also visible, as they reflect light and increase their luminosity.

Now, talking about the last subcategory, reflection nebulae don't emit significant amounts of visible light, but being close friends to stars, they borrow their light by reflecting it.

Similar nebulae not near to stars, being not illuminated, may not emit visible radiation but can be detected as opaque clouds that block light. Also, know best as dark nebulae. Lastly, the main example of diffuse nebula is the Carina nebula.

Moving on, the second type is planetary nebulae. They are remnants of the final stage of stellar evolution for lower-mass stars.

Evolved asymptotic giant branch stars normally tend to expel their outer layers due to strong stellar winds, forming gaseous shells, while as we know leaving the core in the middle in from of a white dwarf.

Just so you know, hot white dwarfs' radiation excited the expelled gases, producing emission nebulae found in star formation regions.

Again, H II regions, as hydrogen's mostly ionized. Yet, planetary nebulae are denser and more compact than others.

Planetary nebulae were given their name by the first astronomical observers who were initially unable to distinguish them from planets, and also for those who simply confused them.

Our Sun could spawn a planetary or diffuse nebula in twelve billion years, even though Andromeda will end us.

A lovely one to look is the Oyster nebula, located in the constellation of Camelopardalis.

The third type is protoplanetary nebula. A PPN nebula is an astronomical object at the short-lived period during a stellar evolution of a star between the late asymptotic giant branch phase and the following planetary nebula phase.

The protoplanetary nebula is energized by the central star, which causes it to emit very strong IR radiation; this process essentially creates a reflection nebula.

Stellar winds coming from the central star, start their journey and shock the shell into an axially symmetric form.

At the same time, they produce a fast moving molecular wind.

To determine if the PPN becomes a PN, is simply defined by the central star's temperature.

The PPN phase stands until temperatures reach thirty thousand kelvin, after which it is hot enough to ionize the surrounding gas.

Very difficult to understand in a brief, so here is a recap in simple words. And an introduction.

As we know, the solar system formed most probably when a massive molecular cloud core collapsed.

That was a dense part of an interstellar medium which contained inside gas and dust, and was freezing cold. When the collapse occurred, it was either spontaneous or was triggered by a supernova.

As the cloud core collapse, most of the mass fell to the centre, and the remaining bits formed a rotationally supported disk, or protoplanetary nebula.

A great example of a PPN, is found in the Auriga constellation, and named the Westbrook nebula, discovered by William E., you guessed it, Westbrook.

PART XIXb – BUT ORION AND WESTBROOK?

I chose these two as they are very fascinating, but millions of nebulae exist, so when you feel inspired, search for as many as you want.

So, the Westbrook Nebula. We just mentioned it a minute or day, whatever, prior and is obviously a protoplanetary nebula.

It's also known as CRL 618, and is located in Auriga. It's formed by a star that has passed through the red giant phase and has a ceased nuclear fusion at its core.

This star is placed in the middle of the nebula. And it ejects gas and dust at very high, I mean very high, velocities.

It was named as we know after Westbrook, after his death when he was just twenty-six years old in 1975.

So you should have understood that they take a lot of time to form, because star death is a very long process, and we previously discussed about stars.

Before moving on, keep note of this facts. The star always evolves, and become hotter. This means that it emits even more ultraviolet light, implying the power requirements to light up the gas will be satisfied.

And remember, that PPNs are short-lived phenomenon, meaning that the probability that there are many found in a single period of time is rare.

The Orion nebula. Now, it's not made by Oreo joining onions, but is a very interesting astronomical body.

Another enormous cloud of gas and dust, which includes associations of stars, ionized gas and many reflection nebulae at once.

This nebula is part of a bigger complex called Orion Molecular Cloud Complex, or OMCC if you like shortening stuff.

It's just thirteen hundred ly from Earth, and long forty ly in diameter. This nebula is sort of a hospital, because it's were many stars actually are given birth.

As I will further explain, it's the brightest nebula we see and maybe know. This means you can find it. Search the three stars of the Orion's belt, or below, and the nebula is halfway down the sword, and will appear an odd star.

That's it. Now, why is it red than? It gets its reddish blue colour from hydrogen, that's being energized by radiation from new-born stars nearby.

The function of the colours, exists and it's black and white simple. Red areas emit strong light, while blue regions reflect radiation that comes from hot O-type stars, which we mentioned.

And to know why it's called partnership with onion company and Oreo, it actually comes from Greek mythology, and was the name of a hunter.

As I mentioned, there's the Orion's belt. Also known as the Three Kings or Three Sisters, is an asterism in the constellation Orion.

Three bright stars are there, called Alnitak, Alnilam and Mintaka. So to know more about this whole belt and nebula, let's explore each body one by one, or just three by, three?

Alnitak. Simple, triple star system. Eastern end of Orion's belt and far away, as you subtract forty from thirteen hundred light years.

We find Alnitak B which is a B-type star that orbits Alnitak A every thousand and a half "Gregorian" years.

The primary Alnitak A is a close binary, comprising Alnitak Aa, a blue supergiant. And we are not joking when we say blue supergiant. It's sort of twenty times or more bigger in every dimension compared to our huge Sun.

In the night sky, it helps Orion's belt to be seen, as is the brightest O-type star we see.

Alnilam is the second body, as is just "one", being another supergiant. It is the twenty-ninth brightest star in the sky. You might think the Sun is bright, as that being that number on the list makes you nothing.

Wrong. Wrong. Wrong. Alnilam is "three hundred and seventy-five thousand" times more luminous than our Sun.

And is not the most luminous. It has a very important place in the Milky Way in this field.

Now, before mentioning exact stars, let's talk about the faintest of all, Mintaka. It's ninety thousand times more luminous than our Sun, and is a double star. These two stars orbit around every six days.

And again, before talking about those, you must know the difference between luminosity and brightness, like the one with velocity and speed.

Brightness is the rate at which the star's radiated energy reaches an observer on Earth, while luminosity is how much it emits. It's like having something pure, pursuing an experiment and expecting the result to have the same mass.

Essentially, we see how bright a star is, a star emit light and therefore is luminous.

So, talking about the Orion's belt, we start with four stars. We have Mintaka, Alnitak and Alnilam as the main representatives, but the real stars lie at the centre.

We have Rigel, Bellatrix, Saiph and Betelgeuse.

Rigel's another blue supergiant, eighty times bigger in diameter than our Sun. It's the brightest star in the constellation of Orion. So is big and bright.

Yet we have no precise idea of the exact apparent brightness, as it varies and is considered an Alpha Cygni-type Star. Rigel also has brothers, with Rigel B and Rigel C surrounding it.

Bellatrix is a double giant star, and again another variable found in the same constellation. Its name means female warrior from Latin. Bellatrix is bigger than our Sun, but not as much as others. In a few million years Bellatrix will become an orange giant, and eventually a white dwarf.

Saiph is a white bluish supergiant that is massive, way bigger than the Sun. This giant has exhausted its hydrogen supplies, and evolved as a consequence.

Saiph's name derives from Arabic meaning sword of the giant. All the seven names I told you have meaning like that.

Now, there is a lot to say about stars, but for now let's stay to the important bits.

Last on the list, Betelgeuse. It's the second most luminous in the Orion constellation, and tenth brightest in the night sky. Its name also comes from Arabic, meaning the giant or the "one who's at the centre", or the giant's shoulder.

So what is it? A distinctly reddish and semi-regular variable star, which is classified as a red supergiant of spectral type M1-2. Is one of the largest stars visible to the naked eye.

To understand how big it is, well if it was our Sun for a second, it would easily swallow or engulf Mercury, Venus, Earth, Mars and Jupiter.

PART XX – HOW MANY CONSTELLATIONS EXIST?

Yes, the nebulae category surely took a lot of reading, and that's nothing compared to what we know.

But now, constellations. They obviously are group of stars that form a pattern. They are easily recognizable, and help people when they are lost.

In total, there are many of them, but eighty-eight are official.

As we know, the creative team of names have decided on naming the group of stars after object, animals and people.

As you probably know, they are invisible, meaning they are not connected, and they are simply stars.

The probability to see them really depends on the day, month, season, location and properties of each.

And before telling you more, they do not move, at least often, and it's the Earth which moves.

So let's talk like professionals now.

Some constellations are circumpolar, thus they are never found below the horizon when seen from a location on Earth. And they can be seen in the night sky throughout the year.

There are nine constellations that we know which are circumpolar. These are found in the Northern Hemisphere.

Auriga, Camelopardalis, Cassiopeia, Draco, Cepheus, Perseus, Ursa Major, Lynx and Ursa Minor.

Draco is the eight largest constellation, and its name comes from Latin, meaning dragon. It was discovered by the astronomer Ptolemy in the second century.

It features, like all, so many stars, but the majors are: Eltanin, Aldibain, Rastaban, Altais, Aldhibah and Batentaban Borealis.

Moving on, the largest constellation is called Hydra, and the smallest is Crux. Smaller patterns of stars are not constellations, rather known as asterisms.

PART XXa – WHAT ARE ASTERISMS?

In observational astronomy, an asterism is a popularly known pattern or group of stars that can be seen in the night sky, just like constellations.

So the real difference between those, is that constellations are considered to be formally-known areas of the sky with all their celestial bodies found within, while asterisms are visually obvious collection of stars.

The truth about this, nobody knows. This distinction between terms remains inconsistent, which varies among many astronomers.

Essentially, what could be seen as difference, even though it's a human definition, is asterisms don't have official, at least yet, boundaries and are therefore a general concept.

Another way of explaining this, is to think of them as an informal group of stars, normally found within the area of a "dead" constellation, or crossing the boundaries of other two.

Now, asterisms typically range from simple shapes of just a few stars to more complex collections of many stars, covering large portions of the sky or less.

They tend to be all the same brightness approximately. The asterism called The Ploug, the Big Dipper and other comprises the seven brightest stars in the Ursa Major constellation.

The Summer Triangle is an astronomical asterism in the Northern Hemisphere, with three defining vertices called Altair, Vega and Deneb. Each of this is the brightest star of their constellation.

PART XXI – WHAT ABOUT MUSLIM ASTRONOMY?

Arabic Area, or more specifically Islamic astronomy, has developed such a long history of discoveries.

This normally tends to be within the Islamic Golden Age, and normally noted down in Arabic.

Now, these accomplishments started in the Middle East, Central Asia, India, Far East, North Africa and other places.

Islamic astronomy played an important role during the revival of Byzantine and European astronomy, following the loss of knowledge in the Medieval age.

Discoveries related, still today, many stars in the sky, like Aldebaran, Deneb and Altair are still referred today with their Arabic names. This also applies to many astronomical terms.

A famous "Arabic" astronomer, actually Persian, was Abd- al-Rahman al-Sufi, which we heard a few times. He lived in the 900s, and had many names for which he was called after.

He was also one of the nine famous Muslim astronomers. His main job, knowing where he was from, was to translate and expand the previous Greek astronomical works, for example by Ptolemy.

One thing he noticed early, was the Large Magellanic Cloud. And the people discovering it, found out only in the 16th century. So he was well ahead.

Another example, he discovered the Andromeda galaxy in 964 AD. He initially described it as a small cloud, however that was not his fault.

During the same year, he published his works in the Book of Fixed Stars. He dedicated it to the local ruler, and included some pretty tasty information.

It included about forty-eight constellations. That was huge, considering how many were found a lot later.

PART XXII – THEN GREEK ANSTRONOMERS?

Two important Greek figures, Hesiod and Homer, were much influenced by the myths of Mesopotamia and Phoenicia, thanks to the people like sailors who went for maritime commerce and some to live or work.

And they spread the word. It was also spread by Babylonians and Arameans, who went to Lefkandi during the Orientalizing Period for the same reason.

They were transported later with no choice by the Assyrian army from the reign of Babylonia.

Now, it's a long story and maybe not as interesting. So let's see the context behind it.

Skipping ahead, there were many hints which led some fascinating "news". So, some references to stars and constellations appear throughout.

In also occurred in the Iliad and the Odyssey, were many astronomical phenomena and solar eclipses were introduced.

This included many, but the most famous are Ursa Major, Orion and Sirius, the star.

Following on, the word planet came from the Greek, meaning wanderer, as astronomical bodies moved around.

Indeed, the five planets visible from Earth, were all named after Greek names, like Zeus, Cronus, Aphrodite, Ares and Hermes. Even the Sun and the Moon.

They indicated that Venus was two bodies, which later on was proven to be just the planet by Pythagoras.

In classical Greece, astronomy was essentially mathematics. So one day, a young mathematician "taught" by Plato, came up with a two-sphere model, yet geocentric, which divided the cosmos into two regions, a spherical Earth and heaven.

Now jokes aside, even though heaven and Earth might have some problems, he gave it to Plato to publish it in the Republic, where he reported the eight circles, including seven planets and fixed stars.

In addition, Greece also included things like sirens, so it is important to focus on the stuff we know so far. Not saying they could exist or not, as we have no real idea what's in the sea.

Next up, Hellenistic astronomy. The Eudoxan system seen before had its flaws in motion.

But here the history is quite difficult, so the important part for this part occurred in the 2nd century BC, where Hipparchus aware of the Babylonians' accuracy, sort of stole some information.

For the Sun, he used a simple eccentric, based on observations of the equinoxes. He explained the changes for the speed and seasons' length difference.

While for the Moon, he used a deferent and epicycle model. He could not do much more and criticized others for unprofessional models.

He compiled a star catalogue of his own. He observed a nova.

Then, the enlightenment came. Aristarchus of Samos, another important astronomer created the first heliocentric model, arranging the Sun at the centre, finally.

His ideas clearly were not seen as revolutionary, indeed not much has been found, and we only know some names of his students.

He also wrote a book, On the Sizes and Distances of the Sun and Moon. This was the only intact piece of work fully recovered.

Many considered Hipparchus to be among the best, but Claudius Ptolemy, a genius mathematician whom works are still considered phenomenal.

He wrote many books, and The Almagest, very influential in Western astronomy, explained how to predict the behaviour of each planet.

This was due to the mathematical tool invented by him called the equant.

Now, some had problems with this, yet he continued. It was later disposed by the heliocentric and Tychonic system.

PART XXIII – WHO WAS NICOLAUS COPERNICUS?

Nicolas Copernicus was a Renaissance polymath, mathematician, astronomer and much more, who re-invented the model of the universe with the Sun at its centre.

Now, he developed this based on the model of Aristarchus of Samos, who did this early model way before.

In fact, Copernicus was born in 1473, and lived seventy years.

Indeed, in 1543, before dying, his publication of the book De revolutionibus orbium coelestium triggered science with the Copernican revolution.

He made some observations of Mercury, Venus and Mars, other four with Jupiter and Saturn.

He also saw many events, like an eclipse of the Moon in 1500.

After being a secretary and physician, he started studying what we want to know more about, the heliocentric model.

He made a little commentary about this system called Commentarious. This described some ideas and seven main assumptions, after which began studying more about the subject.

In 1532, when the work on the previous book, with a long Latin name, were nearly finished, he did not want to spread the news yet. However, a year later, Johann Albrecht Widmannstetter contacted Pope Clement VII. He liked them, and wrote to Copernicus.

Essentially, the recap was, spread the word, please, we are dying to find out.

And the situation got complex. Still, his book was published, someday, with the fear of heavy criticism.

Before that, in 1593, Georg Joachim Rheticus, arrived in Frombork. He was asked to learn more about some astronomers, and became Copernicus' pupil.

He wrote the book Narratio Prima, presenting briefly Copernicus' theory. And after pushing hard

Copernicus to publish the book, he agreed and was sent away to be printed.

He died on the day where he was presented with the last printed pages of the book.

PART XXIV – DID ROMANS KNOW THE SAME?

Even though Ptolemy lived in the Roman Greece, they still knew a lot about astronomy, at least they did what was possible at the time.

So, the Romans knew seven of the celestial bodies in the sky.

Indeed, they could see, with their naked eye, the moon, the Sun, Mercury, Venus, Mars, Jupiter.

But as we know, these names all come from Roman Gods. The other two bluish planets and Saturn were not yet discovered by the empire.

And with the introduction of the Roman calendar, they used it to guess the possible future events.

Now, they did not know they were actually planets, so they called them wandering stars.

The most part were done by Hipparchus, Ptolemy and others.

PART XXV – ANYTHING WE ACTUALLY MISSED?

If you thought they were bright, well that's wrong. Planets as we should already know have no important light sources.

The only reason they can be seen at some hours of the day, is due to the Sun, which shoots light and is reflected by the nearest bodies.

Obviously, the larger the astronomical body, the more light that's reflected. And no, the universe is not a solid body, that's why it's black like the universe!

And the reason for which stars are sometimes less brighter compared to planets, is simply because in the solar system, planets are closer to Earth than stars.

For example, what is the hottest body in the whole universe? Not that simple, as it's the dead star found in the centre of the Red Spider Nebula. It has a surface temperature of a hundred and forty thousand degrees Celsius. And it is a white dwarf.

In 1974, NASA created several space suits, so who cares. Well, at the time they cost twelve million dollars, and adjusted for inflation today, they would be over a hundred and fifty million dollar worth.

Because the Sun is ninety-eight percent of the mass of the solar system, we could fit over a million Earths inside the Sun, with a "smashed" shape.

As we know, trees are still a thing on Earth, but did you know there are more trees than stars in the milky way. There are three trillion stars and just less than four hundred billion in this galaxy.

If Mars was the next planet we would be living in, which is impossible, than romantic couples would face a problem. Would we still like the sunset.

Yes, the sunset is not as we know, because it's blue. According to NASA, it's fine dust that makes the blue near the Sun's part of the sky more visible.

And if you want to become the first humans with over a quadrillion dollars, well check this out. A planet, called 55 Cancri e, is most likely made of graphite and diamond. And mining it just a piece could be over billions of dollars.

Have you ever been sad about finding out that ice cream is not hot. Well, it is. This planet, an exoplanet, called Gliese 436 b, is thirty-three light years away.

Essentially, it's composed of water elements. However, it's surface surpasses three hundred degrees Celsius, so it would heat up. But, the ice still stands and remains because of pressure. Fun, right?

Furthermore, we know that there is a big gas cloud made from alcohol, and it's a thousand times larger in diameter than the solar system's!

And to end the book, here's another cart full of curiosities.

Did you know, probably not, that Jupiter has hurricane three times the size of Earth. So as said, never think that something makes sense in life.

This "Great Red Spot" has been torturing Jupiter for so many centuries.

Second fact, even cooler, is that the solar system is actually enclosed in a swarm of icy bodies known as Oort cloud, and as a whole. We cannot see this because the particles are so far away, that our eye has a too small field of view and zoom capability.

Before 1980, we only knew well enough about our planets. From then, thousands of planets outside of the solar system have been discovered, called exoplanets. There may be over thirty thousand planets within one light year from us.

Just to keep remembering stuff, the Universe is made up of normal matter, dark matter and dark energy.

Another possible fact, includes the chance of our Universe ending with the Big Freeze.

And if you are not satisfied, which I know you are, the Big Rip might occur. Essentially, this will happen if the dark energy increases without limit.

` Have you ever thought that the Universe has a centre, well you are wrong.

Also, how cool is this! Neutron stars, which we talked about, can spin over six hundred times per second.

And talking about humanity for a second here, a fully fledge out NASA space suit will set you back about twelve million dollars.

Want something even more stupid and absurd. Well it's very impossible and unlikely to thrust a fart in space. Such devastating news.

For the next big astronauts growing up, just know that you cannot jump off the Moon's surface, unless you have a speed of two kilometres per second.

So, you probably understood that outside of this tiny beautiful problematic world, you'll find a piece of everything in every corner, even though there are no corners!

It's amazing to believe how astronomy and cosmology were first studied such a long time ago, and it's like we're just sitting here proving them right.

Still though, always believe that knowledge must be found outside, out and about, but must be kept inside, that's the reason for the name.

Every time you watch the sky, or nature itself, think again and elaborate information. Look at the colours, combinations, patterns and reasons for why things are generally as they are, at least for now.

www.ingramcontent.com/pod-product-compliance
Lightning Source LLC
Chambersburg PA
CBHW070110230526
45472CB00004B/1204